YOUR KNOWLEDGE HAS VALUE

General Combination Theorem and Selected Combinations

Deapon Biswas

Bibliographic information published by the German National Library:

The German National Library lists this publication in the National Bibliography; detailed bibliographic data are available on the Internet at http://dnb.dnb.de.

ISBN: 9783389011836
This book is also available as an ebook.

© GRIN Publishing GmbH
Trappentreustraße 1
80339 München

Print and binding: Books on Demand GmbH, Norderstedt, Germany
Printed on acid-free paper from responsible sources.

The present work has been carefully prepared. Nevertheless, authors and publishers do not incur liability for the correctness of information, notes, links and advice as well as any printing errors.

GRIN web shop: https://www.grin.com/document/1464423

General Combination Theorem and Selected Combinations

Deapon Biswas

Transport Officer, Private Concern, Chattogram, Bangladesh

Abstract

So for combinations are discussed with different theorems in algebra. In this chapter I apply assembly analysis to get the theorems easy and memorable. After assembly analysis applied there becomes a lot of new theorems and all the theorems get a new face by summation methods.

Keywords

Combination space, combination member, combination component, identified combination, combination event, general combination theorem, combination distribution, combination expansion, selected combinations.

Article Outline

1. Introduction
2. Preliminaries
3. Combination Space
4. Combination Member
5. Combination Component
6. Identified Combination
7. Combination Event
8. General Combination Theorem
9. Combination Distribution
10. Combination Expansion
11. Selected Combinations
12. Conclusion

1. Introduction

We have a full idea about combination. It indicates the outcome of a random experiment. That is combination is the selection of M different components taken V at a time what is called usually a random experiment where order is not taken into account and repetitions are not allowed.

2. Preliminaries

As a preliminary it may be stated the following theorem

$$C\binom{N}{V} = \frac{(N-V+1)(N-V+2)(N-V+3)...N}{V!} \; ; \; V \leq N \quad\text{——————— (1)}$$

3. Combination Space

A combination space is a set of all possible combinations (outcomes) of an experiment from a parent assembly A where the outcomes do not take order of the components into account. Let a combination space contains T possible outcomes then the combination space denoted by $C\{^A_V\}$ is

$$C\{^A_V\} = \{C_1, C_2, C_3, \cdots, C_t, \cdots, C_T\} \quad\text{——————— (2)}$$

$$\text{where, } V = 1, 2, 3, \ldots\ldots, N$$
$$N = \text{Parent component number.}$$

Example 1: Set a combination space of the experiment " 4 letters a, b, c, and d select 3 at a time ".

Solution: We have given $A = (a, b, c, d)$ and $V = 3$.

Thus the combination space is

$$C\left\{\begin{matrix}(a, b, c, d)\\3\end{matrix}\right\} = \{(a, b, c), (a, b, d), (a, c, d), (b, c, d)\}.$$

4. Combination Member

A combination member is an element of the combination space (1) usually denoted by C_t ; $t = 1, 2, 3, \ldots\ldots, T$; is

$$C_t = (C_{t1}, C_{t2}, C_{t3}, \ldots, C_{tv}, \ldots, C_{tV}) \quad\text{——————— (2)}$$

The combination member (2) contains V combination components. For everyday use we use the word "combination" to mean combination member.

Example 2: Find the combinations C_1, C_3 and C_4 of the example 1.

Solution: The desired combinations are $C_1 = (a, b, c)$, $C_3 = (a, c, d)$ and $C_4 = (b, c, d)$.

5. Combination Component

It is an element of the combination member (2) usually denoted by C_{tv}; $t = 1, 2, 3,\ldots\ldots, T$; $v = 1, 2, 3,\ldots\ldots, V$ which states the combination component takes v^{th} place in the tth combination member in the combination space (1).

Example 3: Find the combination components C_{12}, C_{23}, C_{33} and C_{42} of the example 1.

Solution: The desired components are $C_{12} = b$, $C_{23} = d$, $C_{33} = d$ and $C_{42} = c$.

Theorem 1: The number of combination of N different components taken V at a time denoted by $C\binom{N}{V}$ is

$$C\binom{N}{V} = \frac{(N-V+1)(N-V+2)(N-V+3)\ldots N}{V!} \ ; \ V \leq N \qquad\text{_____ (3)}$$

Using the summation method we get (3) as

$$C\binom{N}{V} = \sum_{k_1=1}^{(N-V+1)} \sum_{k_2=k_1+1}^{(N-V+2)} \sum_{k_3=k_2+1}^{(N-V+3)} \cdots \sum_{k_v=k_{v-1}+1}^{N} C \qquad\text{_____ (4)}$$

Proof: We know in a combination a change in order does not make a new combination. Thus it is enough to record the selections which arise from a particular direction, i.e., to say the selections of three letters from first four English alphabet a, b, c and d where order is not taken into account are abc, abd, acd and bcd. Here each selection shows the pattern that the letters placed continually from left to right of the alphabet and no need to the selections as cba, dab etc. Let we here V places to be filled with out of N components that all different. First arrange the N components as an O. P. A where first component, second component, third components and so on N^{th} component. The O. P. A. can be designed as

$$A = (\ \square\ ,\square\ ,\ \square\ ,\ldots,\ \square\ ,\ldots,\ \square\ )$$

$\qquad 1^{st} \quad 2^{nd} \quad 3^{rd} \qquad n^{th} \qquad N^{th}$

Fig. 1

We fill the first place with one of the first (N−V+1) components of the O. P. A. that begins from first component and ends in $(N-V+1)^{th}$ component of the row. If the remaining (V−1) components whose any one put in the first place we must fill the remaining $(V-1)$ places with using at least one of the first $(N-V+1)$ components and thus the direction breaks. Now k_1 states an index that indicates a component of the row taken by first place of a combination takes the interval

$$1 \leq k_1 \leq (N-V+1) \quad\text{————————————} \quad (5)$$

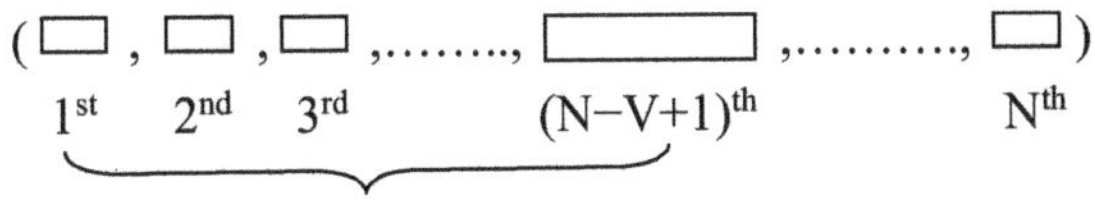

first place takes the component

Fig. 2

Thus the number of events of combinations characterizing first components denoted by S_1, is the number of indexes held by the interval (5) i.e.,

$$S_1 = (N-V+1) - 1 + 1 = (N-V+1) \quad\text{————————} \quad (6)$$

Using the summation method we get (6) as

$$S_1 = \sum_{k_1=1}^{(N-V+1)} C = (N-V+1) \times 1 = (N-V+1) \quad\text{————} \quad (7)$$

where C is a constant quantity taking unit value.

The second place fill with one of the $\{(N-V+2) - (k_1+1) + 1\}$ components of the O. P. A. that begins from $(k_1+1)^{th}$ component and ends in $(N-V+2)^{th}$ component of the O. P. A. Now k_2 states an index that indicates a component of the O. P. A. taken by second place of the combination holds the interval

$$(k_1+1) \leq k_2 \leq (N-V+2) \quad\text{———————} \quad (8)$$

Clearly when

$$k_1 = 1 \quad \text{then} \quad 2 \leq k_2 \leq (N-V+2) \qquad ; \quad (N-V+1)$$
$$k_1 = 2 \quad \text{then} \quad 3 \leq k_2 \leq (N-V+2) \qquad ; \quad (N-V)$$
$$k_1 = 3 \quad \text{then} \quad 4 \leq k_2 \leq (N-V+2) \qquad ; \quad (N-V-1)$$
$$\vdots$$
$$k_1 = (N-V+1) \quad \text{then} \quad (N-V+2) \leq k_2 \leq (N-V+2) \qquad ; \quad 1$$

The numbers after semicolons gives the number of indexes held by the corresponding intervals.

first place takes the component

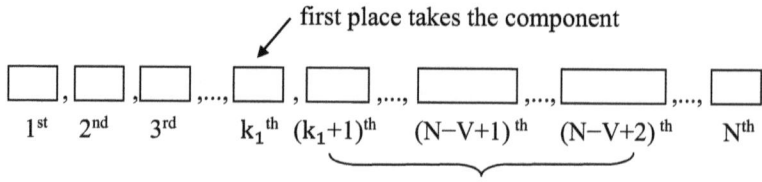

1st 2nd 3rd $k_1{}^{th}$ $(k_1+1)^{th}$ $(N-V+1)^{th}$ $(N-V+2)^{th}$ N^{th}

second place takes the components

Fig. 3

Thus the number of events of combinations characterizing second components denoted by S_2 is the sum of numbers of indexes held by the interval (8) i.e.,

$$S_2 = (N-V+1) + (N-V) + (N-V-1) + \ldots\ldots + 1$$
$$= \frac{(N-V+1)(N-V+2)}{2!} \hspace{3cm} (9)$$

Using the summation method we get (13.4.11) as

$$S_2 = \sum_{k_1=1}^{(N-V+1)} \sum_{k_2=k_1+1}^{(N-V+2)} C \hspace{3cm} (10)$$

Again the third place fill with one of the $\{(N-V+3)-(k_2+1)+1\}$ components of the O. P. A. that begins from $(k_2+1)^{th}$ component and ends in $(N-V+3)^{th}$ component of the O. P. A. Now k_3 states an index that indicates a component of the O. P. A. taken by third place of the combination holds the interval

$$(k_2+1) \leq k_3 \leq (N-V+3) \hspace{3cm} (11)$$

Clearly when

$k_1 = 1$ and $k_2 = 2$ then $3 \leq k_3 \leq (N-V+3)$; $(N-V+1)$

$k_1 = 1$ and $k_2 = 3$ then $4 \leq k_3 \leq (N-V+3)$; $(N-V)$

$k_1 = 1$ and $k_2 = 4$ then $5 \leq k_3 \leq (N-V+3)$; $(N-V-1)$

$\vdots$

$k_1 = 1$ and $k_2 = (N-V+2)$ then $(N-V+3) \leq k_3 \leq (N-V+3)$; 1

$k_1 = 2$ and $k_2 = 3$ then $4 \leq k_3 \leq (N-V+3)$; $(N-V)$

$k_1 = 2$ and $k_2 = 4$ then $5 \leq k_3 \leq (N-V+3)$; $(N-V-1)$

$k_1 = 2$ and $k_2 = 5$ then $6 \leq k_3 \leq (N-V+3)$; $(N-V-2)$

$\vdots$

$k_1 = 2$ and $k_2 = (N-V+2)$ then $(N-V+3) \leq k_3 \leq (N-V+3)$; 1

Similarity when

$k_1 = (N-V+1)$ and $k_2 = (N-V+2)$ then $(N-V+3) \leq k_3 \leq (N-V+3)$; 1

The numbers after semicolons give the number of indexes held by the corresponding intervals. Thus the number of events characterizing third components denoted by S_3 is the sum of numbers of indexes held by the interval (11) i.e.,

$$S_3 = (N-V+1) + (N-V) + (N-V-1) + \ldots + 1 + (N-V) + (N-V-1) + \ldots + 1$$
$$+ \ldots + 1$$

$$= \frac{1}{2}(N-V+1)(N-V+2) + \frac{1}{2}(N-V)(N-V+1) + \ldots + \frac{1}{2} \cdot 1.2 \quad \text{——— (12)}$$

Putting $(N-V+1) = a$, we get the general term of the series, (12) as

$$\frac{1}{2}a(a+1) = \frac{1}{2}(a^2+a)$$

Now the sum of the series (12) is

$$S_3 = \frac{1}{2}\left\{\frac{a(a+1)(2a+1)}{6} + \frac{a(a+1)}{2}\right\} = \frac{a(a+1)(a+2)}{3!}$$

Again putting $a = (N-V+1)$, we get the sum as the number of events characterizing third components i.e.,

$$S_3 = \frac{(N-V+1)(N-V+2)(N-V+3)}{3!} \quad \text{——— (13)}$$

Using the summation method we get (13) as

$$S_3 = \sum_{k_1=1}^{(N-V+1)} \sum_{k_2=k_1+1}^{(N-V+2)} \sum_{k_3=k_2+1}^{(N-V+3)} C \quad \text{——— (14)}$$

Continuing this process we get the Vth place fill with one of the $\{N - (k_{V-1} + 1) + 1\}$ components of the O. P. A. that begins from $(k_{V-1}+1)^{th}$ component and ends in N^{th} component of the row, i.e., k_V states an index that indicates a component of the row taken by V^{th} place of the combination holds the interval

$$(k_{V-1}+1) \leq k_V \leq N \quad \text{——— (15)}$$

Thus the number of events of combinations characterizing V^{th} components denoted by S_V is the sum of numbers of indexes held by the set of interval (13.4.17) i.e.,

$$S_V = \frac{(N-V+1)(N-V+2)(N-V+3) \ldots N}{V!} \quad \text{——— (16)}$$

Using the summation method we get (16) as

$$S_V = \sum_{k_1=1}^{(N-V+1)} \sum_{k_2=k_1+1}^{(N-V+2)} \sum_{k_3=k_2+1}^{(N-V+3)} \ldots \sum_{k_V=k_{V-1}+1}^{N} C \quad \text{——— (17)}$$

As the events characterizing V^{th} components are of one combination, so the equations (16) and (17) give the number of combinations of N different components taken V at a time.

Example 4: Find the number of combination of 7 different components taken 4 at a time using the summation method.

Solution: We have given $N = 7$, $V = 4$ and $N-V+1 = 4$.

Now from (4) we get

$$C\binom{7}{4} = \sum_{k_1=1}^{4} \sum_{k_2=k_1+1}^{5} \sum_{k_3=k_2+1}^{6} \sum_{k_4=k_3+1}^{7} C$$

$$= \sum_{k_2=2}^{5} \sum_{k_3=k_2+1}^{6} \sum_{k_4=k_3+1}^{7} C + \sum_{k_2=3}^{5} \sum_{k_3=k_2+1}^{6} \sum_{k_4=k_3+1}^{7} C$$

$$+ \sum_{k_2=4}^{5} \sum_{k_3=k_2+1}^{6} \sum_{k_4=k_3+1}^{7} C + \sum_{k_2=5}^{5} \sum_{k_3=k_2+1}^{6} \sum_{k_4=k_3+1}^{7} C$$

$$= \left(\sum_{k_3=3}^{6} \sum_{k_4=k_3+1}^{7} C + \sum_{k_3=4}^{6} \sum_{k_4=k_3+1}^{7} C + \sum_{k_3=5}^{6} \sum_{k_4=k_3+1}^{7} C + \right.$$
$$\left. \sum_{k_3=6}^{6} \sum_{k_4=k_3=1}^{7} C \right) + \left(\sum_{k_3=4}^{6} \sum_{k_4=k_3+1}^{7} C + \sum_{k_3=5}^{6} \sum_{k_4=k_3+1}^{7} C + \right.$$
$$\left. \sum_{k_3=6}^{6} \sum_{k_4=k_3+1}^{7} C \right) + \left(\sum_{k_3=5}^{6} \sum_{k_4=k_3+1}^{7} C + \sum_{k_3=6}^{6} \sum_{k_4=k_3+1}^{7} C \right) +$$
$$\left(\sum_{k_3=6}^{6} \sum_{k_4=k_3+1}^{7} C \right)$$

$$= \left(\sum_{k_4=4}^{7} C + \sum_{k_4=5}^{7} C + \sum_{k_4=6}^{7} C + \sum_{k_4=7}^{7} C + \sum_{k_4=5}^{7} C + \sum_{k_4=6}^{7} C + \sum_{k_4=7}^{7} C \right.$$
$$\left. + \sum_{k_4=6}^{7} C + \sum_{k_4=7}^{7} C + \sum_{k_4=7}^{7} C \right) + \left(\sum_{k_4=5}^{7} C + \sum_{k_4=6}^{7} C + \sum_{k_4=7}^{7} C + \sum_{k_4=6}^{7} C \right.$$
$$\left. + \sum_{k_4=7}^{7} C + \sum_{k_4=7}^{7} C \right) + \left(\sum_{k_4=6}^{7} C + \sum_{k_4=7}^{7} C + \sum_{k_4=7}^{7} C \right) + \left(\sum_{k_4=7}^{7} C \right)$$

$$= (4C+3C+2C+C+3C+2C+C+2C+C+C)+(3C + 2C + C + 2C + C + C) +$$
$$(2C + C + C) + (C)$$
$$= 20C + 10C + 4C + C$$
$$= 35C$$
$$= 35 \times 1 \quad \text{[taking } C = 1\text{]}$$
$$= 35.$$

Theorem 2: The total number of combinations of N different components denoted by $C\binom{N}{V}_{V \in \mathcal{V}}$ is

$$C\binom{N}{V}_{V \in \mathcal{V}} = \sum_V \frac{(N-V+1)\,(N-V+2)\,(N-V+3)........N}{V!} \qquad (18)$$

$$\text{where}\,, \quad V = 1, 2, 3,.........., N$$

6. Identified Combination

An identified combination is a combination member of V components whose first v components are identified i.e., the first v components have no right to change their places where as the others have. It is denoted by $C_{tV/v}$ i.e.,

$$C_{tV/v} = \left(\underline{C_{t1}, C_{t2}, C_{t3}, \dots, C_{tv}}, \dots, C_{tV}\right) \quad\quad\quad (19)$$

The identified components are to be underlined.

Example 5: Find the identified combinations of the combination member $C_t = (a, b, c, d, e, f)$ whose (i) first component identified, (ii) first two components identified and (iii) first three components identified.

Solution: The identified combinations are (i) $C_{t6/1} = (\underline{a}, b, c, d, e, f)$, (ii) $C_{t6/2} = (\underline{a, b,} c, d, e, f)$ and (iii) $C_{t6/3} = (\underline{a, b, c,} d, e, f)$.

7. Combination Event

It is a special kind of subset of a combination space where the combination members are to be have same first components, same second components, same third components and so on same v^{th} components. Suppose the combination members

$$C_1 = (C_{11}, C_{12}, C_{13}, \dots, C_{1v}, \dots, C_{1V})$$
$$C_2 = (C_{21}, C_{22}, C_{23}, \dots, C_{2v}, \dots, C_{2V})$$
$$C_3 = (C_{31}, C_{32}, C_{33}, \dots, C_{3v}, \dots, C_{3V})$$
$$\vdots$$
$$C_t = (C_{t1}, C_{t2}, C_{t3}, \dots, C_{tv}, \dots, C_{tV})$$

$$\text{where, } C_{11} = C_{21} = C_{31} = \dots = C_{t1}$$
$$C_{12} = C_{22} = C_{32} = \dots = C_{t2}$$
$$C_{13} = C_{23} = C_{33} = \dots = C_{t3}$$
$$\vdots$$
$$C_{1v} = C_{2v} = C_{3v} = \dots = C_{tv}$$

Then the combination event denoted by $C\left\{{A \atop V/^vA}\right\}$ consists of $C_1, C_2, C_3, \dots, C_t$ i.e.,

$$C\left\{{A \atop V/^vA}\right\} = \{C_1, C_2, C_3, \dots, C_t\} \quad\quad\quad (20)$$

Here A is a parent assembly containing N components, V is the number of components occurred in a combination member and $^V A$ is an identified component assembly containing v identified components. The combination event contains any number of combination members and starts with any combination member of $C\begin{Bmatrix} A \\ V \end{Bmatrix}$ contains the condition supports.

Example 6: Find the combination events of the example 1 where identified components are (i) $^1 A = (a)$ and (ii) $^2 A = (a, b)$.

Solution: The combination events are

(i) $C\begin{Bmatrix} (a, b, c, d) \\ 3/(a) \end{Bmatrix} = \{(a, b, c), (a, b, d), (a, c, d)\}.$

(ii) $C\begin{Bmatrix} (a, b, c, d) \\ 3/(a, b) \end{Bmatrix} = \{(a, b, c), (a, b, d)\}.$

Theorem 3: The number of combination of N different components taken V at a time whose first v components are identified, denoted by $C\begin{pmatrix} N \\ V/v \end{pmatrix}$ is

$$C\begin{pmatrix} N \\ V/^V A \end{pmatrix} = \frac{(N-V+v-k_v+1)\,(N-V+v-k_v+2)\ldots\ldots(N-k_v)}{(V-v)!} \underline{\quad\quad} \quad (21)$$

Using the summation method we get (21) as

$$C\begin{pmatrix} N \\ V/^V A \end{pmatrix} = \sum_{k_{v+1}=k_v+1}^{(N-V+v+1)} \sum_{k_{v+2}=k_{v+1}+1}^{(N-V+v+2)} \sum_{k_{v+3}=k_{v+2}+1}^{(N-V+v+3)} \cdots \sum_{k_v=k_{v-1}+1}^{N} C$$

$$\underline{\quad\quad\quad\quad\quad} \quad (22)$$

where C is a constant quantity taking unit value. The combinations are ordered by the relation that of the O. P. A.

Proof: As the combinations are ordered by the relation that of the O. P. A. and there are first v components identified so the last (V−v) components of combinations are taken from last (N−k_v) components of the row. Now the $(v+1)^{th}$ place fill with one of the $\{(N-V+v+1)-(k_v+1)+1\}$ components of the row that begins from $(k_v+1)^{th}$ component and ends in $(N-V+v+1)^{th}$ component of the row. So k_{v+1} states an index that indicates a component of the row taken by $(v+1)^{th}$ place of a combination holds the interval

$$(k_v+1) \le k_{v+1} \le (N-V+v+1) \underline{\quad\quad\quad} \quad (23)$$

As k_v is an index that indicates the identified component taken by v^{th} place of the combination thus k_v does not varies. Thus the number of events of combinations characterizing $(v+1)^{th}$ components whose first v components are identified denoted by

$S_{v+1/v}$ is the number of indexes held by the interval (23) i.e.,

$$S_{v+1/v} = (N-V+ v+1)-(k_v +1) + 1$$
$$= N-V+ v - k_v + 1 \qquad\qquad (24)$$

Using the summation method we get (24) as

$$S_{v+1/v} = \Sigma_{k_{v+1}=k_v+1}^{(N-V+ v+1)} C \qquad\qquad (25)$$

where C is a constant quantity taking unit value. Again the $(v+2)^{th}$ place can be filled with one of the $\{(N-V+ v+2)-(k_{v+1} +1) +1\}$ components that begins from $(k_{v+1} +1)^{th}$ component and ends in $(N-V+v+2)^{th}$ component of the O. P. A. Now k_{v+2} states an index that indicates a component of the row taken by $(v+2)^{th}$ place of the combination holds the interval

$$(k_{v+1} +1) \leq k_{v+2} \leq (N-V+ v +2) \qquad\qquad (26)$$

As k_{v+1} is an indexes that indicates a component of the O. P. A. taken by $(v+1)^{th}$ place of the combination and the component is not identified, so k_{v+1} varies.

Clearly when,

$k_{v+1} = k_v +1$ then $(k_v +2) \leq k_{v+2} \leq (N-V+v+2)$; $(N-V+ v-k_v +1)$

$k_{v+1} = k_v +2$ then $(k_v +3) \leq k_{v+2} \leq (N-V+v+2)$; $(N-V+ v-k_v)$

$k_{v+1} = k_v +3$ then $(k_v +4) \leq k_{v+2} \leq (N-V+v+2)$; $(N-V+ v-k_v-1)$

$\vdots$

$k_{v+1} = (N-V+v+1)$ then $(N-V+v+2) \leq k_{v+2} \leq (N-V+v+2)$; 1

The numbers after semicolons give the number of indexes held by the corresponding intervals. Thus the number of events of combinations characterizing $(v+2)^{th}$ components whose first v components are identified, denoted by $S_{v+2/v}$ is the sum of numbers of indexes held by the interval (26) i.e.,

$$S_{v+2/v} = (N-V+ v- k_v +1) + (N-V+ v- k_v) + (N-V+ v-k_v-1) +.....+ 1$$
$$= \frac{(N-V+ v - k_v + 1)(N-V+ v - k_v + 2)}{2!} \qquad\qquad (27)$$

Using the summation method we get (27) as

$$S_{v+2/v} = \Sigma_{k_{v+1}=k_v+1}^{(N-V+ v+1)} \Sigma_{k_{v+2}=k_{v+1}+1}^{(N-V+ v+2)} C \qquad\qquad (28)$$

Again the $(v+3)^{th}$ place can be filled with one of the $\{(N-V+v+3) - (k_{v+2}+1)+1\}$ components that begins from $(k_{v+2}+1)^{th}$ component and ends in $(N-V+v+3)^{th}$ component of the O. P. A. Now k_{v+3} states an index that indicates a component of the O. P. A. taken by $(v+3)^{th}$ place of the combination holds the interval

$$(k_{v+2}+1) \leq k_{v+3} \leq (N-V+v+3) \quad\text{————————(29)}$$

As k_{v+2} is an index that indicates a component of the row taken by $(v+2)^{th}$ place of the combination and the component is not identified so k_{v+2} varies.

Clearly when,

$k_{v+1} = k_v+1$ and $k_{v+2} = k_v+2$

then $(k_v+3) \leq k_{v+3} \leq (N-V+v+3)$; $(N-V+v-k_v+1)$

$k_{v+1} = k_v+1$ and $k_{v+2} = k_v+3$

$\qquad$ then $(k_v+4) \leq k_{v+3} \leq (N-V+v+3)$; $(N-V+v-k_v)$

$k_{v+1} = k_v+1$ and $k_{v+2} = k_v+4$

$\qquad$ then $(k_v+5) \leq k_{v+3} \leq (N-V+v+3)$; $(N-V+v-k_v-1)$

$\qquad \vdots$

$k_{v+1} = k_v+1$ and $k_{v+2} = (N-V+v+2)$

$\qquad$ then $(N-V+v+3) \leq k_{v+3} \leq (N-V+v+3)$; 1

$k_{v+1} = k_v+2$ and $k_{v+2} = k_v+3$

$\qquad$ then $(k_v+4) \leq k_{v+3} \leq (N-V+v+3)$; $(N-V+v-k_v)$

$k_{v+1} = k_v+2$ and $k_{v+2} = k_v+4$

$\qquad$ then $(k_v+5) \leq k_{v+3} \leq (N-V+v+3)$; $(N-V+v-k_v-1)$

$k_{v+1} = k_v+2$ and $k_{v+2} = k_v+5$

$\qquad$ then $(k_v+6) \leq k_{v+3} \leq (N-V+v+3)$; $(N-V+v-k_v-2)$

$\qquad \vdots$

$k_{v+1} = k_v+2$ and $k_{v+2} = (N-V+v+2)$

$\qquad$ then $(N-V+v+3) \leq k_{v+3} \leq (N-V+v+3)$; 1

Similarly when

$k_{v+1} = (N-V+v+1)$ and $k_{v+2} = (N-V+v+2)$

$\qquad$ then $(N-V+v+3) \leq k_{v+3} \leq (N-V+v+3)$; 1

The numbers after semicolons give the number of indexes held by the corresponding intervals. Thus the number of events characterizing $(v+3)^{th}$

components, whose first v components are identified, denoted by $S_{v+3/v}$ is the sum of numbers of indexes held by the interval (29) i.e.,

$$S_{v+3/v} = (N-V+v-k_v+1) + (N-V+v-k_v) + \ldots\ldots + 1$$
$$+ (N-V+v-k_v) + (N-V+v-k_v-1) + (N-V+v-k_v-2)$$
$$+\ldots\ldots + 1 + \ldots. + 1$$
$$= \frac{1}{2}(N-V+v-k_v+1)(N-V+v-k_v+2) + \frac{1}{2}(N-V+v-k_v)$$
$$(N-V+v-k_v+1) + \ldots. + \frac{1}{2}1.2 \quad\text{------------}\quad (30)$$

Putting $(N-V+v-k_v+1) = a$ we get the general term of the series (30) as

$$\frac{1}{2}a(a+1) = \frac{1}{2}(a^2+a)$$

Now the sum of the series (30) is

$$S_{v+3/v} = \frac{1}{2}\left\{\frac{a(a+1)(2a+1)}{6} + \frac{a(a+1)}{2}\right\} = \frac{a(a+1)(a+2)}{3!}$$

Again putting $a = (N-V+v-k_v+1)$ we get the sum as the number of events characterizing $(v+3)^{th}$ components i.e.,

$$S_{v+3/v} = \frac{(N-V+v-k_v+1)(N-V+v-k_v+2)(N-V+v-k_v+3)}{3!} \quad\text{------}\quad (31)$$

Using the summation method we get (31) as

$$S_{v+3/v} = \sum_{k_{v+1}=k_v+1}^{(N-V+v+1)} \sum_{k_{v+2}=k_{v+1}+1}^{(N-V+v+2)} \sum_{k_{v+3}=k_{v+2}+1}^{(N-V+v+3)} C \quad\text{------------}\quad (32)$$

Continuing this process we get the vth place filled with one of the $\{N-(k_{V-1}+1)+1\}$ components of the O. P. A. that begins from $(k_{V-1}+1)^{th}$ component and ends in N^{th} component of the O. P. A. Now k_V states an index that indicates a component of the O. P. A. taken by V^{th} place of the combination, holds the interval

$$(k_{V-1}+1) \leq k_V \leq N \qquad [\text{See } (15)]$$

So the number of events of combinations characterizing v^{th} components whose first v components are identified, denoted by $S_{V/v}$ is

$$S_{V/v} = \frac{(N-V+v-k_v+1)(N-V+v-k_v+2)(N-V+v-k_v+3)\ldots\ldots(N-k_v)}{(V-v)!} \quad\text{---}\quad (33)$$

Using the summation method we get (33) as

$$S_{V/v} = \sum_{k_{v+1}=k_v+1}^{(N-V+v+1)} \sum_{k_{v+2}=k_{v+1}+1}^{(N-V+v+2)} \sum_{k_{v+3}=k_{v+2}+1}^{(N-V+v+3)} \cdots \sum_{k_V=k_{V-1}+1}^{N} C$$
$$\text{-----------}\quad (34)$$

where C is a constant quantity taking unit value. As the events characterizing v^{th} components whose first v components identified are of one combination. So the equation (33) or (34) give the number of combinations of N different components taken V at a time.

Example 7: How many ways can a selection of 4 be chosen from a O. P. A. of 8 letters A, B, C, D, E, F, G and H such that

(i) they are starting with A

(ii) there is found the letter A

(iii) they are starting with D

(iv) there is found the letter D

(v) they are starting with C and E as first and second letters respectively

(vi) there are found the letters C and E.

The selections are ordered by the relation that of the row of 8 letters.

Solution: Let the O. P. A. is ordered as

$$A \to B \to C \to D \to E \to F \to G \to H$$

We are given $N = 8$ and $V = 4$.

(i) Here A is the first letter of the given O. P. A. and we fill the first place with the letter A. So, $k_1 = 1$.

The lower and upper limits of k_2 are (k_1+1) and $(N-V+2)$ i.e., 2 and 6 respectively. Now from equation (21) we get

$$C\binom{8}{4/(A)} = \frac{5 \times 6 \times 7}{3!} = 35.$$

Again from equation (22) we get

$$C\binom{8}{4/(A)} = \sum_{k_2=2}^{6} \sum_{k_3=k_2+1}^{7} \sum_{k_4=k_3+1}^{8} C$$

$$= \sum_{k_3=3}^{7} \sum_{k_4=k_3+1}^{8} C + \sum_{k_3=4}^{7} \sum_{k_4=k_3+1}^{8} C + \sum_{k_3=5}^{7} \sum_{k_4=k_3+1}^{8} C$$

$$+ \sum_{k_3=6}^{7} \sum_{k_4=k_3+1}^{8} C + \sum_{k_3=7}^{7} \sum_{k_4=k_3+1}^{8} C$$

$$= \left(\sum_{k_4=4}^{8} C + \sum_{k_4=5}^{8} C + \sum_{k_4=6}^{8} C + \sum_{k_4=7}^{8} C + \sum_{k_4=8}^{8} C \right) + \left(\sum_{k_4=5}^{8} C + \sum_{k_4=6}^{8} C + \sum_{k_4=7}^{8} C + \sum_{k_4=8}^{8} C \right) + \left(\sum_{k_4=7}^{8} C + \sum_{k_4=8}^{8} C \right) + \left(\sum_{k_4=8}^{8} C \right)$$

$$= (5C+4C+3C+2C+C) + (4C+3C+2C+C) + (3C+2C+C) + (2C+C) + (C)$$

$$= 35 C$$

$$= 35 \times 1 \quad [\text{taking } C = 1]$$

$$= 35.$$

(ii) Since A is the first letter of the given O. P. A. and the selections are ordered by the relation that of the given O. P. A. thus no selections of 4 we left on the causation of the letter A found in other places of selections except first place. Hence the desired number is the same as (i) i.e., 35.

(iii) Here D is the 4th letter in the given O. P. A. and we fill the first place with D. Thus the question follows the theorem 13.4.8. Now $k_1 = 4$ and the lower and upper limits of k_2 are $(k_1 + 1)$ and $(N-V+2)$ i.e., 5 and 6 respectively. Hence we get from equation (21) the desired number is

$$C\binom{8}{4/(D)} = \frac{2 \times 3 \times 4}{3!} = 4.$$

Again from equation (22) we get the desired number is

$$C\binom{8}{4/(D)} = \sum_{k_2=5}^{6} \sum_{k_3=k_2+1}^{7} \sum_{k_4=k_3+1}^{8} C$$
$$= \sum_{k_3=6}^{7} \sum_{k_4=k_3+1}^{8} C + \sum_{k_3=7}^{7} \sum_{k_4=k_3+1}^{8} C$$
$$= \left(\sum_{k_4=7}^{8} C + \sum_{k_4=8}^{8} C \right) + \left(\sum_{k_4=8}^{8} C \right)$$
$$= (2C + C) + (C)$$
$$= 3C + C$$
$$= 4C$$
$$= 4 \times 1 \quad [\text{taking } C = 1]$$
$$= 4.$$

(iv) Since D is the fourth so we left lots of selection of 4 on the causation of the letter D found in other places of selections except first place. So the desired number is the same as the number of selections of 4 from a row of 8 letters. When consider D as the first letter in the O. P. A. Then the number of selections starting D is

$$\sum_{k_2=2}^{6} \sum_{k_3=k_2+1}^{7} \sum_{k_4=k_3+1}^{8} C = 35.$$

which is the number of selections where found the letter D.

(v) Since C is the third letter in the O. P. A. Thus $k_1 = 3$.

The lower and upper limits of k_2 are $(k_1 + 1)$ and $(N-V+2)$ i.e., 4 and 6 respectively. As the second letter of the selections is given which is E, the fifth letter in the row thus $k_2 = 5$.

Now the lower and upper limits of k_3 are $(k_2 + 1)$ and $(N-V+3)$ i.e., 6 and 7 respectively. Hence the desired number is the number that states the combinations whose first two components are identified, i.e.,

$$C\binom{8}{4/(C, E)} = \sum_{k_3=6}^{7} \sum_{k_4=k_3+1}^{8} C$$
$$= 2C + C$$
$$= 3C$$

$$= 3 \times 1 \qquad [\text{taking } C = 1]$$
$$= 3.$$

Now for your use the selections are CEFG, CEFH and CEGH.

Note: They will get same number of selections starting with A and E, B and E, D and E as first and second letters respectively.

(vi) Since there are found the letters C and E, so we may consider C and E as the first and second letters respectively in the given O. P. A. So $k_1 = 1$, $k_2 = 2$.

Now the lower and upper limits of k_3 are (k_2+1) and $(N-V+3)$ i.e., 3 and 7 respectively. Thus the desired number is

$$C\binom{8}{4/2} = \sum_{k_3=3}^{7} \sum_{k_4=k_3+1}^{8} C = 15.$$

Corollary 1: The number of combinations of N different components taken V at a time, whose first v components are identified, is the same as the number of combinations of $(N-k_v)$ different components taken $(V-v)$ at a time i.e.,

$$C\binom{N}{V/^vA} = C\binom{N-k_v}{V-v} \qquad \underline{\hspace{3cm}} \quad (35)$$

8. General Combination Theorem

The number of combinations in a combination space of N different components taken V at a time where there occurred Y particular components taken from $U \leq V$ particular components (limited size of components) that occurred in the parent component assembly of N components denoted by $C\binom{N\ U}{V\ Y}$ is

$$C\binom{N\ U}{V\ Y} = C\binom{U}{Y}C\binom{N-U}{V-Y} \qquad \underline{\hspace{3cm}} \quad (40)$$

$$\text{where, } N \geq V$$
$$U \leq V$$
$$U-N+V \leq Y \leq U$$

Y takes only non-negative values.

Proof: Let the parent component assembly is of N components. The combination members are of V components. Now we can select Y components out of U components without restrictions is $C\binom{U}{Y}$. Then for each combination member there remains (V–Y) components to select out of (N–U) components. The number of these selections is $C\binom{N-U}{V-Y}$. Hence finally we get the total number of combinations is $C\binom{U}{Y}C\binom{N-U}{V-Y}$.

Example 8: Let the parent component assembly M = (A, B, C, D, E). Find the number of combinations of the following and then write the combinations.

(i) $C\binom{5\ 3}{3\ 3}$, (ii) $C\binom{5\ 3}{3\ 2}$, (iii) $C\binom{5\ 3}{3\ 1}$, (iv) $C\binom{5\ 2}{3\ 2}$, (v) $C\binom{5\ 2}{3\ 1}$,

(vi) $C\binom{5\ 2}{3\ 0}$,

(vii) $C\binom{5\ 1}{3\ 1}$, (viii) $C\binom{5\ 1}{3\ 0}$, (ix) $C\binom{5\ 3}{4\ 3}$, (x) $C\binom{5\ 3}{4\ 2}$, (xi)

$C\binom{5\ 2}{4\ 2}$, (xii) $C\binom{5\ 2}{4\ 1}$

For (i) to (iii) the components of limited size is (A, B, C)

For (iv) to (vi) the components of limited size is (A, B)

For (vii) to (viii) the components of limited size is (A)

For (ix) to (x) the components of limited size is (A, B, C)

For (xi) to (xii) the components of limited size is (A, B).

Solution:

(i) $C\binom{5\ 3}{3\ 3} = C\binom{3}{3}C\binom{5-3}{3-3} = 1 \times 1 = 1$.

Now the combination is (A, B, C).

(ii) $C\binom{5\ 3}{3\ 2} = C\binom{3}{2}C\binom{5-3}{3-2} = 3 \times 2 = 6$.

Now the combinations are (A, B, D), (A, B, E), (A, C, D), (A, C, E), (B, C, D), (B, C, E).

(iii) $C\binom{5\ 3}{3\ 1} = C\binom{3}{1}C\binom{5-3}{3-1} = 3 \times 1 = 3$.

Now the combinations are (A, D, E), (B, D, E), (C, D, E).

(iv) $C\binom{5\ 2}{3\ 2} = C\binom{2}{2}C\binom{5-2}{3-2} = 1 \times 3 = 3$.

Now the combinations are (A, B, C), (A, B, D), (A, B, E).

(v) $C\binom{5\ \ 2}{3\ \ 1} = C\binom{2}{1}C\binom{5-2}{3-1} = 2 \times 3 = 6.$

Now the combinations are (A, C, D), (A, C, E), (A, D, E), (B, C, D), (B, C, E), (B, D, E).

(vi) $C\binom{5\ \ 2}{3\ \ 0} = C\binom{2}{0}C\binom{5-2}{3-0} = 1 \times 1 = 1.$

Now the combination is (C, D, E).

(vii) $C\binom{5\ \ 1}{3\ \ 1} = C\binom{1}{1}C\binom{5-1}{3-1} = 1 \times 6 = 6.$

Now the combinations are (A, B, C), (A, B, D), (A, B, E), (A, C, D), (A, C, E), (A, D, E).

(viii) $C\binom{5\ \ 1}{3\ \ 0} = C\binom{1}{0}C\binom{5-1}{3-0} = 1 \times 4 = 4.$

Now the combinations are (B, C, D), (B, C, E), (B, D, E), (C, D, E).

(ix) $C\binom{5\ \ 3}{4\ \ 3} = C\binom{3}{3}C\binom{5-3}{4-3} = 1 \times 2 = 2.$

Now the combinations are (A, B, C, D), (A, B, C, E).

(x) $C\binom{5\ \ 3}{4\ \ 2} = C\binom{3}{2}C\binom{5-3}{4-2} = 3 \times 1 = 3.$

Now the combinations are (A, B, D, E), (A, C, D, E), (B, C, D, E).

(xi) $C\binom{5\ \ 2}{4\ \ 2} = C\binom{2}{2}C\binom{5-2}{4-2} = 1 \times 3 = 3.$

Now the combinations are (A, B, C, D), (A, B, C, E), (A, B, D, E).

(xii) $C\binom{5\ \ 2}{4\ \ 1} = C\binom{2}{1}C\binom{5-2}{4-1} = 2 \times 1 = 2.$

Now the combinations are (A ,C, D, E), (B, C, D, E).

The present discussion enables us the probability law in which variables of frequency distribution got from a combination space. We get a theoretical discrete distribution specially for this chapter called combination distribution. However this distribution is the same as that of hyper geometric distribution beside some exceptions. The exceptions are to be stated at due time.

9. Combination Distribution

A random variable Y is said to follow combination distribution if it assumes only non-negative values and its probability mass function is given by

$$P(Y) = C(Y; N, U, V) = \frac{c\binom{U}{Y}c\binom{N-U}{V-Y}}{c\binom{N}{V}} \; ; \; U-N+V \le Y \le U$$

$$= 0 \; ; \text{ otherwise} \quad \underline{\hspace{3cm}} (41)$$

The three independent finite constants N, U and V are known as the parameters of this distribution. Combination distribution is a discrete distribution as Y can take only the non-negative values under the interval U−N+V ≤ Y ≤ U. Any variable which follows combination distribution is known as combination variate and denoted by the symbol Y ~ C (N, U, V).

Remark: It should be noted that

$$\sum_Y C(Y; N, U, V) = \sum_Y \frac{c\binom{U}{Y} c\binom{N-U}{V-Y}}{c\binom{N}{V}} = 1$$

$$\Rightarrow \sum_Y c\binom{U}{Y} c\binom{N-U}{V-Y} = c\binom{N}{V} \quad \underline{\hspace{3cm}} (42)$$

Example 9: Let the parent component assembly M = (A, B, C, D, E, F, G, H). Find the probability of combinations taken 4 at a time in which they have

(i) first 3 components

(ii) exactly any 2 of first 3 components

(iii) exactly any 1 of first 3 components.

Solution: We have given the parameters N = 8, U = 3 and V = 4.

(i) The probability of Y = 3 is given by

$$C(3; 8, 3, 4) = \frac{c\binom{3}{3}c\binom{8-3}{4-3}}{c\binom{8}{4}} = \frac{5}{70} = 0.071$$

(ii) The probability of Y = 2 is given by

$$C(2; 8, 3, 4) - \frac{c\binom{3}{2}c\binom{8-3}{4-2}}{c\binom{8}{4}} = \frac{30}{70} = 0.428$$

(iii) The probability of Y = 1 is given by

$$C(1; 8, 3, 4) = \frac{c\binom{3}{1}c\binom{8-3}{4-1}}{c\binom{8}{4}} = \frac{30}{70} = 0.428.$$

9.1 Moments:

The first four moments about origin of combination distribution are obtained as follows:

$$\mu_1' = E(Y) = \sum_Y Y\, C(Y\,;\,N,U,V) = \sum_Y Y\, C\binom{U}{Y}\, C\binom{N-U}{V-Y}/C\binom{N}{V}$$

$$= \sum_Y Y \frac{U!}{Y!(U-Y)!} C\binom{N-U}{V-Y}/C\binom{N}{V}$$

$$= \frac{U}{C\binom{N}{V}} \sum_Y C\binom{U-1}{Y-1} C\binom{N-U}{V-Y}$$

$$= \frac{U}{C\binom{N}{V}} \sum_Y C\binom{U-1}{Y-1} C\binom{N-1-U+1}{V-1-Y+1}$$

$$= \frac{U}{C\binom{N}{V}} C\binom{N-1}{V-1} = \frac{UV}{N}.$$

Thus the mean of the combination distribution is $= \frac{UV}{N}$.

$$\mu_2' = E(Y^2) = \sum_Y Y^2\, C(Y\,;\,N,U,V) = \sum_Y Y^2 C\binom{U}{Y}\, C\binom{N-U}{V-Y}/C\binom{N}{V}$$

$$= \sum_Y \{Y(Y-1)+Y\}\, C\binom{U}{Y}\, C\binom{N-U}{V-Y}/C\binom{N}{V}$$

$$= \frac{U(U-1)}{C\binom{N}{V}} \sum_Y \left\{1 + \frac{1}{Y-1}\right\} C\binom{U-2}{Y-2} C\binom{N-U}{V-Y}$$

$$= \frac{U(U-1)}{C\binom{N}{V}} \left[\sum_Y C\binom{U-2}{Y-2} C\binom{N-U}{V-Y} + \sum_Y \frac{1}{Y-1} C\binom{U-2}{Y-2} C\binom{N-U}{V-Y}\right]$$

$$= \frac{U(U-1)}{C\binom{N}{V}} \sum_Y C\binom{U-2}{Y-2} C\binom{N-2-U+2}{V-2-Y+2}$$

$$\qquad\qquad + \frac{U}{C\binom{N}{V}} \sum_Y C\binom{U-1}{Y-1} C\binom{N-1-U+1}{V-1-Y+1}$$

$$= \frac{U(U-1)}{C\binom{N}{V}} \times C\binom{N-2}{V-2} + \frac{U}{C\binom{N}{V}} \times C\binom{N-1}{V-1}$$

$$= \frac{UV(U-1)(V-1)}{N(N-1)} + \frac{UV}{N}.$$

$$\mu_3' = E(Y^3) = \sum_Y Y^3\, C(Y\,;\,N,U,V) = \sum_Y Y^3 C\binom{U}{Y}\, C\binom{N-U}{V-Y}/C\binom{N}{V}$$

$$= \sum_Y \{Y(Y-1)(Y-2)+3Y(Y-1)+Y\}\, C\binom{U}{Y}\, C\binom{N-U}{V-Y}/C\binom{N}{V}$$

$$= \frac{U(U-1)(U-2)}{C\binom{N}{V}} \sum_Y \left\{1 + \frac{3}{(Y-2)} + \frac{1}{(Y-1)(Y-2)}\right\} C\binom{U-3}{Y-3} C\binom{N-U}{V-Y}$$

$$= \frac{U(U-1)(U-2)}{C\binom{N}{V}} C\binom{N-3}{V-3} + \frac{3U(U-1)}{C\binom{N}{V}} C\binom{N-2}{V-2} + \frac{U}{C\binom{N}{V}} C\binom{N-1}{V-1}$$

$$= \frac{UV(U-1)(U-2)(V-1)(V-2)}{N(N-1)(N-2)} + \frac{3UV(U-1)(V-1)}{N(N-1)} + \frac{UV}{N}.$$

$$\mu_4' = E(Y^4) = \sum_Y Y^4\, C(Y\,;\,N,U,V) = \sum_Y Y^4 C\binom{U}{Y}\, C\binom{N-U}{V-Y}/C\binom{N}{V}$$

$$= \sum_Y \{Y(Y-1)(Y-2)(Y-3)+6Y(Y-1)(Y-2)+7Y(Y-1)+Y\}$$

$$\times C\binom{U}{Y} C\binom{N-U}{V-Y} \Big/ C\binom{N}{V}$$

$$= \frac{U(U-1)(U-2)(U-3)}{C\binom{N}{V}} \sum_Y \left\{ 1 + \frac{6}{(Y-3)} + \frac{7}{(Y-2)(Y-3)} + \frac{1}{(Y-1)(Y-2)(Y-3)} \right\}$$

$$\times C\binom{U-4}{Y-4} C\binom{N-U}{V-Y}$$

$$= \frac{U(U-1)(U-2)(U-3)}{C\binom{N}{V}} C\binom{N-4}{V-4} + \frac{6U(U-1)(U-2)}{C\binom{N}{V}} C\binom{N-3}{V-3}$$

$$+ \frac{7U(U-1)}{C\binom{N}{V}} C\binom{N-2}{V-2} + \frac{U}{C\binom{N}{V}} C\binom{N-1}{V-1}$$

$$= \frac{UV(U-1)(U-2)(U-3)(V-1)(V-2)(V-3)}{N(N-1)(N-2)(N-3)} + \frac{6UV(U-1)(U-2)(V-1)(V-2)}{N(N-1)(N-2)}$$

$$+ \frac{7UV(U-1)(V-1)}{N(N-1)} + \frac{UV}{N}.$$

Now variance $\mu_2 = \mu_2' - \mu_1'^2$

$$= \frac{UV(U-1)(V-1)}{N(N-1)} + \frac{UV}{N} - \frac{U^2V^2}{N^2}$$

$$= \frac{NUV(U-1)(V-1) + NUV(N-1) - U^2V^2(N-1)}{N^2(N-1)}$$

$$= \frac{NU^2V^2 - NU^2V - NUV^2 + NUV + N^2UV - NUV - NU^2V^2 + U^2V^2}{N^2(N-1)}$$

$$= \frac{-NU^2V - NUV^2 + N^2UV - NU^2V^2 + U^2V^2}{N^2(N-1)}$$

$$= \frac{UV(N-U)(N-V)}{N^2(N-1)}.$$

9.2: Factorial moments of this distribution:

The kth factorial moment of this distribution is obtained as follows:

$$\mu'_{(k)} = E\{Y^{(k)}\} = \sum_Y Y^{(k)} \frac{C\binom{U}{Y} C\binom{N-U}{V-Y}}{C\binom{N}{V}}$$

$$= U^{(k)} \sum_Y \frac{C\binom{U-k}{Y-k} C\binom{N-U}{V-Y}}{C\binom{N}{V}} = U^{(k)} \frac{C\binom{N-k}{V-k}}{C\binom{N}{V}} = \frac{U^{(k)}V^{(k)}}{N^{(k)}}.$$

Now, $\mu'_{(1)} = E\{Y^{(1)}\} = \frac{U^{(1)}V^{(1)}}{N^{(1)}} = \frac{UV}{N}$

$$\mu'_{(2)} = E\{Y^{(2)}\} = \frac{U^{(2)}V^{(2)}}{N^{(2)}} = \frac{U(U-1)V(V-1)}{N(N-1)} = \frac{UV(U-1)(V-1)}{N(N-1)}$$

$$\mu'_{(3)} = E\{Y^{(3)}\} = \frac{U^{(3)}V^{(3)}}{N^{(3)}} = \frac{U(U-1)(U-2)V(V-1)(V-2)}{N(N-1)(N-2)}$$

$$= \frac{UV(U-1)(U-2)(V-1)(V-2)}{N(N-1)(N-2)}.$$

9.3: Mode of this distribution:

We have, $\dfrac{C(Y\,;N,U,V)}{f(Y-1\,;N\,,U\,,V)} = \dfrac{c\binom{U}{Y}c\binom{N-U}{V-Y}/c\binom{N}{V}}{c\binom{U}{Y-1}c\binom{N-U}{V-Y+1}/c\binom{N}{V}}$

i.e., $= \dfrac{P(Y)}{P(Y-1)} = \dfrac{UV - UY - VY + V^2 - 2Y + U + V + 1}{NY - UY - VY + Y^2}$

Now if $\dfrac{P(Y)}{P(Y-1)} > 1$

then, $UV - UY - VY + V^2 - 2Y + U + V + 1 > NY - UY - VY + Y^2$

or, $Y < \dfrac{(U+1)(V+1)}{(N+2)}$

and if $\dfrac{P(Y)}{P(Y-1)} < 1$

then, $Y > \dfrac{(U+1)(V+1)}{(N+2)}$

We discuss the following two cases:

Case I: when $\dfrac{(U+1)(V+1)}{(N+2)}$ is not an integer.

Let $\dfrac{(U+1)(V+1)}{(N+2)} = t + f$; where t is an integer and f is a fractional such that $0 < f < 1$.

then we get $\dfrac{P(Y)}{P(Y-1)} > 1$; for $Y = k, k+1, k+2, \ldots, t$

$$0 \le k = U - N + V$$

And $\dfrac{P(Y)}{P(Y-1)} < 1$; for $Y = t+1, t+2, t+3, \ldots, U$

$\Rightarrow \dfrac{P(k+1)}{P(k)} > 1$, $\dfrac{P(k+2)}{P(k+1)} > 1$, $\ldots$, $\dfrac{P(t)}{P(t-1)} > 1$

and $\dfrac{P(t+1)}{P(t)} < 1$, $\dfrac{P(t+2)}{P(t+1)} < 1$, $\ldots$, $\dfrac{P(U)}{P(U-1)} < 1$.

Thus $P(k) < P(k+1) < P(k+2) \ldots < P(t-1) < P(t) > P(t+1) > P(t+2) > \ldots > P(U)$.

In this case we have one maximum value and this is t. Thus the modal value is t, the integral part of $\dfrac{(U+1)(V+1)}{(N+2)}$.

Case II: When $\dfrac{(U+1)(V+1)}{(N+2)}$ is an integer.

Let $\dfrac{(U+1)(V+1)}{(N+2)} = t$, an integer then we get

$\dfrac{P(Y)}{P(Y-1)} > 1$; for $Y = k, k+1, k+2, \ldots, t-1$

$$0 \le k = U - N + V$$

$$\frac{P(Y)}{P(Y-1)} = 1 \; ; \; \text{for} \; Y = t$$

and $\frac{P(Y)}{P(Y-1)} < 1 \; ; \; \text{for} \; Y = t+1, t+2, \ldots\ldots, U$

Now proceeding as case I we get

$$P(k) < P(k+1) < P(k+2) \ldots\ldots < P(t-1) = P(t) > P(t+1) >$$
$$P(t+2) > \ldots\ldots > P(U)$$

In this case we get two maximum values i.e., $t-1$ and t. Thus the modal values are $(t-1)$ and t i.e., $\left\{\frac{(U+1)(V+1)}{(N+2)} - 1\right\}$ and $\frac{(U+1)(V+1)}{(N+2)}$.

9.4: Recurrence relation for this distribution:

We have $P(Y) = C\binom{U}{Y} C\binom{N-U}{V-Y} / C\binom{N}{V}$

and $P(Y+1) = C\binom{U}{Y+1} C\binom{N-U}{V-Y-1} / C\binom{N}{V}$

Thus, $\dfrac{P(Y+1)}{P(Y)} = \dfrac{C\binom{U}{Y+1} C\binom{N-U}{V-Y-1} / C\binom{N}{V}}{C\binom{U}{Y} C\binom{N-U}{V-Y} / C\binom{N}{V}} = \dfrac{(U-Y)(V-Y)}{(Y+1)(N-U-V+Y+1)}$

Thus, $P(Y+1) = \left\{\dfrac{(U-Y)(V-Y)}{(Y+1)(N-U-V+Y+1)}\right\} P(Y).$

This is the required recurrence relation.

9.5: Binomial distribution from this distribution:

We have $C(Y; N, U, V) = C\binom{U}{Y} C\binom{N-U}{V-Y} / C\binom{N}{V}$

$$= \frac{U!}{Y!(U-Y)!} \times \frac{(N-U)!}{(V-Y)!(N-U-V+Y)!} \times \frac{V!(N-V)!}{N!}$$

$$= \frac{V!}{Y!(V-Y)!} \times U(U-1)(U-2)\ldots(U-Y+1)$$

$$\times \frac{(N-U)(N-U-1)\ldots(N-U-V+Y+1)}{N(N-1)\ldots(N-V+1)}$$

Dividing by N both numerator and denominator of the right side of the above equation we get

$$C(Y; N, U, V) = C\binom{V}{Y} \times \frac{U}{N}\left(\frac{U}{N} - \frac{1}{N}\right)\ldots\left(\frac{U}{N} - \frac{U-1}{N}\right)$$

$$\times \frac{\left(1 - \frac{U}{N}\right)\left(1 - \frac{U}{N} - \frac{1}{N}\right)\ldots\left(1 - \frac{U}{N} - \frac{V-Y-1}{N}\right)}{\frac{N}{N}\left(1 - \frac{1}{N}\right)\left(1 - \frac{2}{N}\right)\ldots\left(1 - \frac{V-1}{N}\right)}.$$

Taking limit as $N \to \infty$ and putting $\frac{U}{N} = p$

We get

$$\lim_{N \to \infty} C(Y; N, U, V) = C\binom{V}{Y} \underbrace{p.p \dots p}_{Y \text{ times}} \underbrace{(1-p)(1-p) \dots (1-p)}_{(V-Y) \text{ times}}$$

$$= C\binom{V}{Y} p^Y (1-p)^{V-Y}$$
$$= b(Y; p, 1-p)$$
$$= b(Y; p, q); \text{ where } q = 1-p$$

which is the probability function of the binomial distribution with parameter 'p' and 'q'.

Combination distribution is specially designed for a combination space where U is always less than or equal to V in which hyper geometric distribution U does not depend on V. Again the lower limit of the random variable Y is k where $0 \le k = U - N + V$ in which hyper geometric distribution it is 0. The upper limit of it is U, in which hyper geometric distribution it is min(U, V).

Example 10: Let there would made a committee of 4 persons from 6 divisions Dhaka. Chittagong, Rajshahi, Khulna, Sylhet and Barishal in Bangladesh selecting a person. How many probability of getting
    (i) exactly 3 persons from Dhaka , Chittagong and Rajshahi
    (ii) any 2 persons from Dhaka , Chittagong and Rajshahi
    (iii) exactly 2 persons from Dhaka and Chittagong.
Solution: (i) We have given N = 6, U = 3, V = 4 and Y = 3.
Thus the number of favorable cases is

$$C\binom{6 \quad 3}{4 \quad 3} = C\binom{3}{3} C\binom{6-3}{4-3} = 3.$$

But the number of possible cases is $C\binom{6}{4} = 15$.

Thus the probability of getting Y = 3 is

$$C(3 ; 6, 3, 4) = \frac{3}{15} = 0.2$$

(ii) We have given N = 6, U = 3, V = 4 and Y = 2.
Thus the number of favorable cases is

$$C\binom{6 \quad 3}{4 \quad 2} = C\binom{3}{2} C\binom{6-3}{4-2} = 9.$$

But the number of possible cases is $C\binom{6}{4} = 15$.

Thus the probability of getting Y = 2 is

$C(2 ; 6, 3, 4) = \dfrac{9}{15} = 0.6$

(iii) We have given N = 6, U = 2, V = 4 and Y = 2.

Thus the number of favorable cases is

$$C\begin{pmatrix} 6 & 2 \\ 4 & 2 \end{pmatrix} = C\begin{pmatrix} 2 \\ 2 \end{pmatrix} C\begin{pmatrix} 6-2 \\ 4-2 \end{pmatrix} = 6.$$

But the number of possible cases is $C\begin{pmatrix} 6 \\ 4 \end{pmatrix} = 15$.

Thus the probability of getting Y = 2 is

$C(2 ; 6, 2, 4) = \dfrac{6}{15} = 0.4$

10. Combination Expansion

The number of combinations in a combination space of N different components taken V at a time is expended as

$$C\binom{N}{V} = C\binom{U}{0}C\binom{N-U}{V} + C\binom{U}{1}C\binom{N-U}{V-1} + C\binom{U}{2}C\binom{N-U}{V-2}$$

$$+......+ C\binom{U}{y}C\binom{N-U}{V-y} +......+ C\binom{U}{U}C\binom{N-U}{V-U}\text{——— (43)}$$

where y is a value of a random variable Y and U is the limited size of the random variable Y.

11. Selected Combinations

We have ideas of getting the number of selections of N different components taken V at a time. This is a matter when the parent assembly has some alike components. The combination theorem for an assembly of N components that has some alike components is large different from the theorem 1. Next theorem introduces the fact when the components are not all different. Let it has N_1 components alike of one kind, N_2 components alike of another kind, N_3 components alike of a third kind and so on for a hth kind then there called first kind of components occupies from first place, second kind of components occupies from $(N_1+1)^{th}$ place, third kind of components occupies from $(N_1+ N_2+1)^{th}$ place and so on h^{th} kind of components occupies from $(N_1+ N_2+ N_3+.......+ N_{h-1} +1)^{th}$ place in the O. P. A. where $N_1 + N_2 + N_3+.......+ N_h$ = N. Obviously in this investigation we omit the other places.

Theorem 4: The number of combination of N components that not all different and there are N_1 components alike of one kind, N_2 components alike of another kind, N_3 components alike of a third kind and so on for a hth kind, taken V at a time, denoted by $C^*\binom{N}{V}$ is

$$C^*\binom{N}{V} = \sum_{g=1}^{S_{V-1}^*} S_{V.g}^* \qquad\qquad (44)$$

Using the summation method we get (44) as

$$C^*\binom{N}{V} = \sum_{k_1^* \in K_1^*} \sum_{k_2^* \in K_2^*} \sum_{k_3^* \in K_3^*} \cdots \sum_{k_V^* \in K_V^*} C \qquad\qquad (45)$$

where C is a constant quantity taking unit value.

Proof: Let we have N components that not all different. Suppose it has N_1 components alike of one kind, N_2 components alike of another kind, N_3 components alike of a third kind and so on N_h components alike of a hth kind then we have k_1, the index that indicates a component of the O. P. A. taken by first place of a combination, as

$$1 \le k_1 \le (N-V+1) \qquad\qquad [\text{See (5)}]$$

and k_1^*, the index which indicates the component of the O. P. A. which is not alike of previous components indicated by previous indexes of the interval (5) as

$$1 \le k_1^* \le \text{till } (N-V+1) \qquad\qquad (46)$$

Clearly the selected indexes are i.e., k_1^* takes

$1, (N_1 +1), (N_1 + N_2 +1), \ldots\ldots, \text{till } (N-V+1)$

Thus the assembly of assemblies of selected indexes is

$$K_1^* = (K_{11}^*) \qquad\qquad (47)$$

where $K_{11}^* = (1, (N_1 +1), (N_1 + N_2 +1), \ldots\ldots, \text{till } (N-V+1))$

and the number of selected events characterizing first components denoted by S_1^* is the sum of numbers of indexes held by the assembly of assemblies (47) i.e.,

$$S_1^* = S_{11}^* = \sum_{g=1}^{S_0^*} S_{1.g}^* \qquad\qquad (48)$$

Providing $S_0^* = 1$

where $S_{11}^* =$ number of indexes held by the assembly K_{11}^*.

Now using the summation method we get (48) as

$$S_1^* = \sum_{k_1^* \in K_1^*} C \qquad\qquad (49)$$

Again we have k_2, the index that indicates a component of the O. P. A. taken by second place of the combination, as

$(k_1+1) \leq k_2 \leq (N-V+2)$ [See (8)]

and k_2^*, the index which indicates the component of the row which is not alike of previous component indicated by pervious indexes of the interval (8) as

$(k_1^*+1) \leq k_2^* \leq$ till $(N-V+2)$ ——————————— (50)

Clearly the selected indexes are i.e., when

$k_1^* = 1$ then k_2^* takes

$$2, (N_1 +1), (N_1 + N_2 +1), \ldots\ldots, \text{till } (N-V+2)$$

$k_1^* = (N_1+1)$ then k_2^* takes

$$(N_1 +2), (N_1 + N_2 +1), \ldots\ldots, \text{till } (N-V+2)$$

$k_1^* = (N_1 + N_2 +1)$ then k_2^* takes

$$(N_1+ N_2+2), (N_1 + N_2+ N_3+1), \ldots\ldots, \text{till } (N-V+2)$$

and so on.

These series are related by the relation $<$. If to say for first series $2 \geq (N_1+1)$ then (N_1+1) quit and so on for second series, third series etc.

Thus the assembly of assemblies of selected indexes is

$K_2^* = (K_{21}^*, K_{22}^*, K_{23}^*, \ldots, K_{2S_1^*}^*)$ ———————————(51)

where $K_{21}^* = (2, (N_1 +1), (N_1 + N_2 +1), \ldots\ldots, \text{till } (N-V+2))$

        $K_{22}^* = ((N_1 +2), (N_1 + N_2 +1), \ldots\ldots, \text{till } (N-V+2))$

        $K_{23}^* = ((N_1+ N_2+2), (N_1 + N_2+ N_3+1), \ldots\ldots, \text{till}(N-V+2))$

and so on.

Thus the number of selected events characterizing second components, denoted by S_2^* is the sum of numbers of indexes held by the assembly of assemblies (51) i.e.,

$S_2^* = S_{21}^* + S_{22}^* + S_{23}^* + \ldots\ldots\ldots S_1^*$ terms

$\quad = \Sigma_{g=1}^{S_1^*} S_{2.g}^*$ ——————————— (52)

where, $S_{21}^* = $ number of indexes held by the assembly K_{21}^*

        $S_{22}^* = $ number of indexes held by the assembly K_{22}^*

        $S_{23}^* = $ number of indexes held by the assembly K_{23}^*

and so on.

Using the summation method we get (52) as

$S_2^* = \Sigma_{k_1^* \in K_1^*} \Sigma_{k_2^* \in K_2^*} C$ ——————————— (53)

Again we have k_3 , the index that indicates a component of the O. P. A. taken by third place of the combination, as

$(k_2 +1) \le k_3 \le (N-V+3)$ [See (11)]

and k_3^* , the index which indicates a component of the row which is not alike of previous components indicated by pervious indexes of the interval (11) as

$(k_2^* +1) \le k_3^* \le$ till $(N-V+3)$ ——————————————— (54)

Clearly the selected indexes are i.e., when

$k_1^* = 1, \ k_2^* = 2,$ then k_3^* takes

$$3, \ (N_1 +1), \ (N_1 + N_2 +1), \ \ldots\ldots.., \text{ till } (N-V+3)$$

$k_1^* = 1, \ k_2^* = (N_1+1)$ then k_3^* takes

$$(N_1 +2), \ (N_1 + N_2 +1), \ \ldots\ldots.., \text{ till } (N-V+3)$$

$k_1^* = 1, \ k_2^* = (N_1 + N_2 +1)$ then k_3^* takes

$$(N_1+ N_2+2), \ (N_1 + N_2+ N_3+1), \ \ldots\ldots.., \text{ till } (N-V+3)$$

and so on.

These series are related by the relation $<$. If to say for first series $3 \ge$ (N_1+1) or $3 \ge (N_1+ N_2+1)$ etc then (N_1+1) , $(N_1 + N_2 +1)$ etc. quit and so on for second series, third series etc.

Again when,

$k_1^* = (N_1+1) , k_2^* = (N_1+2)$ then k_3^* takes

$$(N_1+3), (N_1+ N_2+1), (N_1 + N_2+ N_3+1), \ \ldots\ldots., \text{ till } (N-V+3)$$

$k_1^* = (N_1+1), \ k_2^* = (N_1+ N_2+1)$ then k_3^* takes

$$(N_1+ N_2+2), \ (N_1+ N_2+ N_3+1), \ \ldots\ldots.., \text{ till } (N-V+3)$$

$k_1^* = (N_1+1), \ k_2^* = (N_1 + N_2+ N_3+1)$ then k_3^* takes

$$(N_1 + N_2+ N_3+2), (N_1 + N_2+ N_3+ N_4+1), \ \ldots., \text{ till } (N-V+3)$$

and so on.

These series are related by the relation $<$. If to say for first series (N_1+3) $\ge (N_1+ N_2+1)$ or $(N_1+3) \ge (N_1+ N_2+ N_3+1)$ etc. then $(N_1+ N_2+1)$, $(N_1+ N_2+ N_3+1)$ etc. quit and so on for second series, third series etc.

Similarly we can find the selected indexes for

$k_1^* = (N_1+ N_2+1), (N_1+ N_2+ N_3+1)$ etc.

Thus the assembly of assemblies of selected indexes is

$K_3^* = (K_{31}^*, K_{32}^*, K_{33}^* , \ \ldots.., K_{3S_2^*}^*)$ ——————————— (55)

where, $K_{31}^* = (3, \ (N_1+1), \ (N_1 + N_2 +1), \ \ldots\ldots.., \text{ till } (N-V+3))$

$\qquad K_{32}^* = ((N_1+2), \ (N_1 + N_2 +1), \ \ldots\ldots.., \text{ till } (N-V+3))$

$$K_{33}^* = ((N_1 + N_2 + 2), (N_1 + N_2 + N_3 + 1), \ldots\ldots, \text{till } (N-V+3))$$

and so on.

Thus the number of selected events characterizing third components, denoted by S_3^* is the sum of numbers of indexes held by the assembly of assemblies (55) i.e.,

$$S_3^* = S_{31}^* + S_{32}^* + S_{33}^* + \ldots\ldots S_2^* \text{ terms}$$

$$= \sum_{g=1}^{S_2^*} S_{3.g}^* \quad\quad\quad\quad\quad\quad\quad (56)$$

where, S^*_{31} = number of indexes held by the assembly K^*_{31}

$\quad\quad\quad S^*_{32}$ = number of indexes held by the assembly K^*_{32}

$\quad\quad\quad S^*_{33}$ = number of indexes held by the assembly K^*_{33}

and so on.

Using the summation method we get (56) as

$$S_3^* = \sum_{k_1^* \in K_1^*} \sum_{k_2^* \in K_2^*} \sum_{k_3^* \in K_3^*} C$$

Proceeding this way we get the number of selected events characterizing V^{th} components denoted by S_V^* is

$$S_V^* = S_{V1}^* S^*_{V1} + S_{V2}^* + S_{V3}^* + \ldots\ldots S_{V-1}^* \text{ terms}$$

$$= \sum_{g=1}^{S_{V-1}^*} S_{V.g}^* \quad\quad\quad\quad\quad\quad\quad (57)$$

Using the summation method we get (57) as

$$S_V^* = \sum_{k_1^* \in K_1^*} \sum_{k_2^* \in K_2^*} \sum_{k_3^* \in K_3^*} \ldots\ldots \sum_{k_V^* \in K_V^*} C \quad\quad\quad\quad (58)$$

As the selected events characterizing V^{th} components are of one combination, so the equation (57) and (58) give the total number of combinations of N components that not all different, taken V at a time.

Example 11: How many 4 sub-tuples can be made by the 9-tuples $(2, 2, 2, 3, 5, 5, 5, 7, 11)$?

Solution: We have given $N = 9$, $V = 4$, $N_1 = 3$, $N_2 = 1$, $N_3 = 3$, $N_4 = 1$, $N_5 = 1$

So, $N-V+1 = 6$.

Now in the interval $1 \leq k_1^* \leq$ till 6, we get the assembly of assemblies of selected indexes as

$$K_1^* = (K_{11}^*)$$

$$\text{where } K_{11}^* = (1, 4, 5)$$

So, $S_1^* = 3$.

Again in the interval $(k_1^* + 1) \leq k_2^* \leq$ till 7, we get the assembly of assemblies of selected indexes as

$$K_2^* = (K_{21}^*, K_{22}^*, K_{23}^*)$$

$$\text{where, } K_{21}^* = (2, 4, 5)$$

$$K_{22}^* = (5)$$

$$K_{23}^* = (6)$$

$$\text{So, } S_2^* = \Sigma_{g=1}^3 S_{2.g}^*$$

$$= S_{21}^* + S_{22}^* + S_{23}^*$$

$$= 3 + 1 + 1$$

$$= 5.$$

In the interval $(k_2^* + 1) \leq k_3^* \leq$ till 8, we get the assembly of assemblies of selected indexes as

$$K_3^* = (K_{31}^*, K_{32}^*, K_{33}^*, K_{34}^*, K_{35}^*)$$

$$\text{where, } K_{31}^* = (3, 4, 5, 8)$$

$$K_{32}^* = (5, 8)$$

$$K_{33}^* = (6, 8)$$

$$K_{34}^* = (6, 8)$$

$$K_{35}^* = (7, 8)$$

$$\text{So, } S_3^* = \Sigma_{g=1}^5 S_{3.g}^*$$

$$= S_{31}^* + S_{32}^* + S_{33}^* + S_{34}^* + S_{35}^*$$

$$= 4 + 2 + 2 + 2 + 2$$

$$= 12.$$

In the interval $(k_3^* + 1) \leq k_4^* \leq$ till 9, we get the assembly of assemblies of selected indexes as

$$K_4^* = (K_{41}^*, K_{42}^*, K_{43}^*, K_{44}^*, K_{45}^* K_{46}^*, K_{47}^*, K_{48}^*, K_{49}^*, K_{410}^*, K_{4.11}^*, K_{4.12}^*)$$

$$\text{where, } K_{41}^* = (4, 5, 8, 9) \quad ; \quad K_{47}^* = (7, 8, 9)$$

$$K_{42}^* = (5, 8, 9) \quad ; \quad K_{48}^* = (9)$$

$$K_{43}^* = (6, 8, 9) \quad ; \quad K_{49}^* = (7, 8, 9)$$

$$K_{44}^* = (9) \quad ; \quad K_{410}^* = (9)$$

$$K_{45}^* = (6, 8, 9) \quad ; \quad K_{411}^* = (8, 9)$$

$$K_{46}^* = (9) \quad ; \quad K_{412}^* = (9)$$

$$\text{So, } S_4^* = \Sigma_{g=1}^{12} S_{4.g}^*$$

$$= S_{41}^* + S_{42}^* + S_{43}^* + S_{44}^* + S_{45}^* + S_{46}^* + S_{47}^* + S_{48}^* + S_{49}^* + S_{410}^*$$

$$+ S_{411}^* + S_{412}^*$$

$$= 4+3+3+1+3+1+3+1+3+1+2+1$$
$$= 26.$$

Thus $C^* \begin{pmatrix} 9 \\ 4 \end{pmatrix} = \sum_{g=1}^{12} S_{4.g}^* = 26.$

Using the summation method we get

$$C^* \begin{pmatrix} 9 \\ 4 \end{pmatrix} = \sum_{k_1^* \in K_1^*} \sum_{k_2^* \in K_2^*} \sum_{k_3^* \in K_3^*} \sum_{k_4^* \in K_4^*} C$$

$$= \sum_{k_2^* \in K_{21}^*} \sum_{k_3^* \in K_3^*} \sum_{k_4^* \in K_4^*} C + \sum_{k_2^* \in K_{22}^*} \sum_{k_3^* \in K_3^*} \sum_{k_4^* \in K_4^*} C$$
$$+ \sum_{k_2^* \in K_{23}^*} \sum_{k_3^* \in K_3^*} \sum_{k_4^* \in K_4^*} C$$

As $K_1^* = (K_{11}^*)$
$$K_{11}^* = (1, 3, 4)$$

$$= (\sum_{k_3^* \in K_{31}^*} \sum_{k_4^* \in K_4^*} C + \sum_{k_3^* \in K_{32}^*} \sum_{k_4^* \in K_4^*} C + \sum_{k_3^* \in K_{33}^*} \sum_{k_4^* \in K_4^*} C) +$$
$$(\sum_{k_3^* \in K_{34}^*} \sum_{k_4^* \in K_4^*} C) + (\sum_{k_3^* \in K_{35}^*} \sum_{k_4^* \in K_4^*} C)$$

As $K_{21}^* = (2, 4, 5)$
$$K_{22}^* = (5)$$
$$K_{23}^* = (6)$$

$$= (\sum_{k_4^* \in K_{41}^*} C + \sum_{k_4^* \in K_{42}^*} C + \sum_{k_4^* \in K_{43}^*} C + \sum_{k_4^* \in K_{44}^*} C) + (\sum_{k_4^* \in K_{45}^*} C$$
$$+ \sum_{k_4^* \in K_{46}^*} C) + (\sum_{k_4^* \in K_{47}^*} C + \sum_{k_4^* \in K_{48}^*} C) + (\sum_{k_4^* \in K_{49}^*} C + \sum_{k_4^* \in K_{410}^*} C)$$
$$+ (\sum_{k_4^* \in K_{411}^*} C + \sum_{k_4^* \in K_{412}^*} C)$$

As $K_{31}^* = (3, 4, 5, 8)$
$$K_{32}^* = (5, 8)$$
$$K_{33}^* = (6, 8)$$
$$K_{34}^* = (6, 8)$$
$$K_{35}^* = (7, 8)$$

$$= (4+3+3+1) + (3+1) + (3+1) + (3 + 1) + (2+1) = 26.$$

As $K_{41}^* = (4, 5, 8, 9)$; $K_{47}^* = (7, 8, 9)$
$K_{42}^* = (5, 8, 9)$; $K_{48}^* = (9)$
$K_{43}^* = (6, 8, 9)$; $K_{49}^* = (7, 8, 9)$
$K_{44}^* = (9)$; $K_{410}^* = (9)$
$K_{45}^* = (6, 8, 9)$; $K_{411}^* = (8, 9)$
$K_{46}^* = (9)$; $K_{412}^* = (9)$

Theorem 5: The number of combination of N components that not all different and there are N_1 components alike of one kind, N_2 components alike of another kind, N_3 components alike of a third kind and so on for a hth kind, taken V at a time, denoted by $C^*\binom{N}{V}$ is

$$C^*\binom{N}{V} = \sum_{g=1}^{S_{V-1}^*} S_{V.g}^* \quad\text{————————————— (59)}$$

Using the summation method we get (59) as

$$C^*\binom{N}{V} = \sum_{k_1^* \in K_1^*} \sum_{k_2^* \in K_2^*} \sum_{k_3^* \in K_3^*} \cdots \sum_{k_V^* \in K_V^*} C \quad\text{——————— (60)}$$

where C is a constant quantity taking unit value.

Proof: Let we have N components that not all different. Suppose it has N_1 components alike of one kind, N_2 components alike of another kind, N_3 components alike of a third kind and so on N_h components alike of a hth kind then we have k_1, the index that indicates a component of the O. P. A. taken by first place of a combination , as

$$1 \le k_1 \le (N-V+1) \qquad\qquad \text{[See (5)]}$$

and k_1^*, the index which indicates the component of the O. P. A. which is not alike of previous components indicated by previous indexes of the interval (5) as

$$1 \le k_1^* \le \text{till } (N-V+1) \quad\text{———————————— (61)}$$

Clearly the selected indexes are i.e., k_1^* takes

1, $(N_1 +1)$, $(N_1 + N_2 +1)$, ..., till $(N-V+1)$

Thus the assembly of assemblies of selected indexes is

$$K_1^* = (K_{11}^*) \quad\text{————————————— (62)}$$

where $K_{11}^* = (1, (N_1 +1), (N_1 + N_2 +1), ..., \text{till } (N-V+1))$

and the number of selected events characterizing first components denoted by S_1^* is the sum of numbers of indexes held by the assembly of assemblies (62) i.e.,

$$S_1^* = S_{11}^* = \sum_{g=1}^{S_0^*} S_{1.g}^* \quad\text{————————————— (63)}$$

Providing $S_0^* = 1$

where S_{11}^* = number of indexes held by the assembly K_{11}^*.

Now using the summation method we get (63) as

$$S_1^* = \sum_{k_1^* \in K_1^*} C \quad\text{————————————— (64)}$$

Again we have k_2 , the index that indicates a component of the O. P. A. taken by second place of the combination, as

$$(k_1 +1) \le k_2 \le (N-V+2) \qquad\qquad \text{[See (8)]}$$

and k_2^*, the index which indicates the component of the row which is not alike of previous component indicated by pervious indexes of the interval (8) as

$$(k_1^* + 1) \leq k_2^* \leq \text{ till } (N-V+2) \hspace{2cm} (65)$$

Clearly the selected indexes are i.e., when

$k_1^* = 1$ then k_2^* takes

$$2, (N_1 + 1), (N_1 + N_2 + 1), \ldots\ldots\ldots, \text{ till } (N-V+2)$$

$k_1^* = (N_1 + 1)$ then k_2^* takes

$$(N_1 + 2), (N_1 + N_2 + 1), \ldots\ldots\ldots, \text{ till } (N-V+2)$$

$k_1^* = (N_1 + N_2 + 1)$ then k_2^* takes

$$(N_1 + N_2 + 2), (N_1 + N_2 + N_3 + 1), \ldots\ldots\ldots, \text{ till } (N-V+2)$$

and so on.

These series are related by the relation $<$. If to say for first series $2 \geq (N_1 + 1)$ then $(N_1 + 1)$ quit and so on for second series, third series etc.

Thus the assembly of assemblies of selected indexes is

$$K_2^* = (K_{21}^*, K_{22}^*, K_{23}^*, \ldots, K_{2S_1^*}^*) \hspace{2cm} (66)$$

where $K_{21}^* = (2, (N_1 + 1), (N_1 + N_2 + 1), \ldots\ldots, \text{ till } (N-V+2))$

$\hspace{1.5cm} K_{22}^* = ((N_1 + 2), (N_1 + N_2 + 1), \ldots\ldots\ldots, \text{ till } (N-V+2))$

$\hspace{1.5cm} K_{23}^* = ((N_1 + N_2 + 2), (N_1 + N_2 + N_3 + 1), \ldots\ldots, \text{ till}(N-V+2))$

and so on.

Thus the number of selected events characterizing second components, denoted by S_2^* is the sum of numbers of indexes held by the assembly of assemblies (66) i.e.,

$$S_2^* = S_{21}^* + S_{22}^* + S_{23}^* + \ldots\ldots\ldots S_1^* \text{ terms}$$

$$= \sum_{g=1}^{S_1^*} S_{2.g}^* \hspace{2cm} (67)$$

where , $S_{21}^* = $ number of indexes held by the assembly K_{21}^*

$\hspace{1.5cm} S_{22}^* = $ number of indexes held by the assembly K_{22}^*

$\hspace{1.5cm} S_{23}^* = $ number of indexes held by the assembly K_{23}^*

and so on.

Using the summation method we get (67) as

$$S_2^* = \sum_{k_1^* \in K_1^*} \sum_{k_2^* \in K_2^*} C \hspace{2cm} (68)$$

Again we have k_3 , the index that indicates a component of the O. P. A. taken by third place of the combination, as

$$(k_2 + 1) \leq k_3 \leq (N-V+3) \hspace{2cm} [\text{See (11)}]$$

and k_3^* , the index which indicates a component of the row which is not alike of previous components indicated by pervious indexes of the interval (11) as

$$(k_2^* +1) \leq k_3^* \leq \text{till} (N-V+3) \quad \text{————————} \quad (69)$$

Clearly the selected indexes are i.e., when

$k_1^* = 1$, $k_2^* = 2$, then k_3^* takes

$$3, (N_1 +1), (N_1 + N_2 +1), \ldots\ldots, \text{till} (N-V+3)$$

$k_1^* = 1$, $k_2^* = (N_1+1)$ then k_3^* takes

$$(N_1 +2), (N_1 + N_2 +1), \ldots\ldots, \text{till} (N-V+3)$$

$k_1^* = 1$, $k_2^* = (N_1 + N_2 +1)$ then k_3^* takes

$$(N_1+ N_2+2), (N_1 + N_2+ N_3+1), \ldots\ldots, \text{till} (N-V+3)$$

and so on.

These series are related by the relation $<$. If to say for first series $3 \geq (N_1+1)$ or $3 \geq (N_1+ N_2+1)$ etc then (N_1+1) , $(N_1 + N_2 +1)$ etc. quit and so on for second series, third series etc.

Again when,

$k_1^* = (N_1+1)$, $k_2^* = (N_1+2)$ then k_3^* takes

$$(N_1+3), (N_1+ N_2+1), (N_1 + N_2+ N_3+1), \ldots\ldots, \text{till} (N-V+3)$$

$k_1^* = (N_1+1)$, $k_2^* = (N_1+ N_2+1)$ then k_3^* takes

$$(N_1+ N_2+2), (N_1 + N_2+ N_3+1), \ldots\ldots, \text{till} (N-V+3)$$

$k_1^* = (N_1+1)$, $k_2^* = (N_1 + N_2+ N_3+1)$ then k_3^* takes

$$(N_1 + N_2+ N_3+2), (N_1 + N_2+ N_3+ N_4+1), \ldots, \text{till} (N-V+3)$$

and so on.

These series are related by the relation $<$. If to say for first series $(N_1+3) \geq (N_1+ N_2+1)$ or $(N_1+3) \geq (N_1+ N_2+ N_3+1)$ etc. then $(N_1+ N_2+1)$, $(N_1+ N_2+ N_3+1)$ etc. quit and so on for second series, third series etc.

Similarly we can find the selected indexes for

$k_1^* = (N_1+ N_2+1), (N_1+ N_2+ N_3+1)$ etc.

Thus the assembly of assemblies of selected indexes is

$$K_3^* = (K_{31}^*, K_{32}^*, K_{33}^*, \ldots.., K_{3S_2^*}^*) \quad \text{————————} \quad (70)$$

where, $K_{31}^* = (3, (N_1+1), (N_1 + N_2 +1), \ldots\ldots, \text{till} (N-V+3))$

$K_{32}^* = ((N_1+2), (N_1 + N_2 +1), \ldots\ldots, \text{till} (N-V+3))$

$K_{33}^* = ((N_1 + N_2+2), (N_1 + N_2+ N_3+1), \ldots\ldots, \text{till} (N-V+3))$

and so on.

Thus the number of selected events characterizing third components, denoted by S_3^* is the sum of numbers of indexes held by the assembly of assemblies (70) i.e.,

$$S_3^* = S_{31}^* + S_{32}^* + S_{33}^* + \ldots\ldots S_2^* \text{ terms}$$

$$= \sum_{g=1}^{S_2^*} S_{3.g}^* \quad\quad\quad\quad\quad\quad\quad\quad\quad (71)$$

where, $S*_{31}$ = number of indexes held by the assembly $K*_{31}$

$\quad\quad\quad S*_{32}$ = number of indexes held by the assembly $K*_{32}$

$\quad\quad\quad S*_{33}$ = number of indexes held by the assembly $K*_{33}$

and so on.

Using the summation method we get (71) as

$$S_3^* = \sum_{k_1^* \in K_1^*} \sum_{k_2^* \in K_2^*} \sum_{k_3^* \in K_3^*} C$$

Proceeding this way we get the number of selected events characterizing V^{th} components denoted by S_V^* is

$$S_V^* = S_{V1}^* S*_{VI} + S_{V2}^* + S_{V3}^* + \ldots\ldots S_{V-1}^* \text{ terms}$$

$$= \sum_{g=1}^{S_{V-1}^*} S_{V.g}^* \quad\quad\quad\quad\quad\quad\quad\quad (72)$$

Using the summation method we get (72) as

$$S_V^* = \sum_{k_1^* \in K_1^*} \sum_{k_2^* \in K_2^*} \sum_{k_3^* \in K_3^*} \ldots \sum_{k_V^* \in K_V^*} C \quad\quad\quad\quad (73)$$

As the selected events characterizing V^{th} components are of one combination, so the equation (72) and (73) give the total number of combinations of N components that not all different, taken V at a time.

Example 12: How many 4 sub-tuples can be made by the 9-tuples (2, 2, 2, 3, 5, 5, 5, 7, 11) ?

Solution: We have given N = 9, V = 4, $N_1 = 3$, $N_2 = 1$, $N_3 = 3$, $N_4 = 1$, $N_5 = 1$

So, $N - V + 1 - 6$.

Now in the interval $1 \le k_1^* \le$ till 6, we get the assembly of assemblies of selected indexes as

$$K_1^* = (K_{11}^*)$$

$\quad\quad$ where $K_{11}^* = (1, 4, 5)$

So, $S_1^* = 3$.

Again in the interval $(k_1^* + 1) \le k_2^* \le$ till 7, we get the assembly of assemblies of selected indexes as

$$K_2^* = (K_{21}^*, K_{22}^*, K_{23}^*)$$

where, $K_{21}^* = (2, 4, 5)$

$K_{22}^* = (5)$

$K_{23}^* = (6)$

So, $S_2^* = \sum_{g=1}^{3} S_{2.g}^*$

$= S_{21}^* + S_{22}^* + S_{23}^*$

$= 3 + 1 + 1$

$= 5.$

In the interval $(k_2^* +1) \leq k_3^* \leq$ till 8, we get the assembly of assemblies of selected indexes as

$K_3^* = (K_{31}^*, K_{32}^*, K_{33}^*, K_{34}^*, K_{35}^*)$

where, $K_{31}^* = (3, 4, 5, 8)$

$K_{32}^* = (5, 8)$

$K_{33}^* = (6, 8)$

$K_{34}^* = (6, 8)$

$K_{35}^* = (7, 8)$

So, $S_3^* = \sum_{g=1}^{5} S_{3.g}^*$

$= S_{31}^* + S_{32}^* + S_{33}^* + S_{34}^* + S_{35}^*$

$= 4 + 2 + 2 + 2 + 2$

$= 12.$

In the interval $(k_3^* +1) \leq k_4^* \leq$ till 9, we get the assembly of assemblies of selected indexes as

$K_4^* = (K_{41}^*, K_{42}^*, K_{43}^*, K_{44}^*, K_{45}^* K_{46}^*, K_{47}^*, K_{48}^*, K_{49}^*, K_{410}^*, K_{4.11}^*, K_{4.12}^*)$

where, $K_{41}^* = (4, 5, 8, 9)$; $K_{47}^* = (7, 8, 9)$

$K_{42}^* = (5, 8, 9)$; $K_{48}^* = (9)$

$K_{43}^* = (6, 8, 9)$; $K_{49}^* = (7, 8, 9)$

$K_{44}^* = (9)$; $K_{410}^* = (9)$

$K_{45}^* = (6, 8, 9)$; $K_{411}^* = (8, 9)$

$K_{46}^* = (9)$; $K_{412}^* = (9)$

So, $S_4^* = \sum_{g=1}^{12} S_{4.g}^*$

$= S_{41}^* + S_{42}^* + S_{43}^* + S_{44}^* + S_{45}^* + S_{46}^* + S_{47}^* + S_{48}^* + S_{49}^* + S_{410}^*$

$\quad + S_{411}^* + S_{412}^*$

$= 4 + 3 + 3 + 1 + 3 + 1 + 3 + 1 + 3 + 1 + 2 + 1$

$= 26.$

Thus $C^* \binom{9}{4} = \sum_{g=1}^{12} S_{4.g}^* = 26.$

Using the summation method we get

$$C^* \binom{9}{4} = \sum_{k_1^* \in K_1^*} \sum_{k_2^* \in K_2^*} \sum_{k_3^* \in K_3^*} \sum_{k_4^* \in K_4^*} C$$

$$= \sum_{k_2^* \in K_{21}^*} \sum_{k_3^* \in K_3^*} \sum_{k_4^* \in K_4^*} C + \sum_{k_2^* \in K_{22}^*} \sum_{k_3^* \in K_3^*} \sum_{k_4^* \in K_4^*} C$$

$$+ \sum_{k_2^* \in K_{23}^*} \sum_{k_3^* \in K_3^*} \sum_{k_4^* \in K_4^*} C$$

As $K_1^* = (K_{11}^*)$

$$K_{11}^* = (1, 3, 4)$$

$$= (\sum_{k_3^* \in K_{31}^*} \sum_{k_4^* \in K_4^*} C + \sum_{k_3^* \in K_{32}^*} \sum_{k_4^* \in K_4^*} C + \sum_{k_3^* \in K_{33}^*} \sum_{k_4^* \in K_4^*} C) +$$

$$(\sum_{k_3^* \in K_{34}^*} \sum_{k_4^* \in K_4^*} C) + (\sum_{k_3^* \in K_{35}^*} \sum_{k_4^* \in K_4^*} C)$$

As $K_{21}^* = (2, 4, 5)$

$$K_{22}^* = (5)$$

$$K_{23}^* = (6)$$

$$= (\sum_{k_4^* \in K_{41}^*} C + \sum_{k_4^* \in K_{42}^*} C + \sum_{k_4^* \in K_{43}^*} C + \sum_{k_4^* \in K_{44}^*} C) + (\sum_{k_4^* \in K_{45}^*} C$$

$$+ \sum_{k_4^* \in K_{46}^*} C) + (\sum_{k_4^* \in K_{47}^*} C + \sum_{k_4^* \in K_{48}^*} C) + (\sum_{k_4^* \in K_{49}^*} C + \sum_{k_4^* \in K_{410}^*} C)$$

$$+ (\sum_{k_4^* \in K_{411}^*} C + \sum_{k_4^* \in K_{412}^*} C)$$

As $K_{31}^* = (3, 4, 5, 8)$

$$K_{32}^* = (5, 8)$$

$$K_{33}^* = (6, 8)$$

$$K_{34}^* = (6, 8)$$

$$K_{35}^* = (7, 8)$$

$$= (4+3+3+1) + (3+1) + (3+1) + (3 + 1) + (2+1) = 26.$$

As $K_{41}^* = (4, 5, 8, 9)$; $K_{47}^* = (7, 8, 9)$

$\quad K_{42}^* = (5, 8, 9)$; $K_{48}^* = (9)$

$\quad K_{43}^* = (6, 8, 9)$; $K_{49}^* = (7, 8, 9)$

$\quad K_{44}^* = (9)$; $K_{410}^* = (9)$

$\quad K_{45}^* = (6, 8, 9)$; $K_{411}^* = (8, 9)$

$\quad K_{46}^* = (9)$; $K_{412}^* = (9)$

Example 13: In the example 12 how many 4-subtuples can be made with 2 and 2 as first and second digits respectively ? 2 and 3 as first and second digits respectively ?

Solution: We have given 2 and 2 as first and second digits respectively. Thus the question requires the number of combinations taken 4 at a time whose first 2 components are identified, then $k_1^* = 1$ and $k_2^* = 2$.

Now in the interval $3 \leq k_3^* \leq$ till 8, we get the assembly of assemblies of selected indexers as

$K_3^* = (K_{31}^*)$ where, $K_{31}^* = (3, 4, 5, 8)$

So, $S_3^* = 4$.

Again in the interval $k_3^* +1 \leq k_4^* \leq$ till 9, we get the assembly of assemblies of selected indexes as

$K_4^* = (K_{41}^*, K_{42}^*, K_{43}^*, K_{44}^*)$ where, $K_{41}^* = (4, 5, 8, 9)$

$$K_{42}^* = (5, 8, 9)$$
$$K_{43}^* = (6, 8, 9)$$
$$K_{44}^* = (9)$$

So, $S_4^* = \sum_{g=1}^{S_3^*} S_{4.g}^*$

$\qquad = S_{41}^* + S_{42}^* + S_{43}^* + S_{44}^*$

$\qquad = 4 + 3 + 3 + 1$

$\qquad = 11.$

Thus $C^* \begin{pmatrix} 9 \\ 4/(2,2) \end{pmatrix} = 11.$

Using the summation method we get

$C^* \begin{pmatrix} 9 \\ 4/(2,2) \end{pmatrix} = \sum_{k_3^* \in K_3^*} \sum_{k_4^* \in K_4^*} C$

$\qquad\qquad = \sum_{k_4^* \in K_{41}^*} C + \sum_{k_4^* \in K_{42}^*} C + \sum_{k_4^* \in K_{43}^*} C + \sum_{k_4^* \in K_{44}^*} C ;$

As $K_3^* = (K_{31}^*)$

$K_{31}^* = (3, 4, 5, 8)$

$\qquad = 4 + 3 + 3 + 1 \quad ;$ $K_{41}^* = (4, 5, 8, 9)$

$\qquad = 11$ $K_{42}^* = (5, 8, 9)$

$\qquad\qquad\qquad\qquad\qquad\qquad K_{43}^* = (6, 8, 9)$

$\qquad\qquad\qquad\qquad\qquad\qquad K_{44}^* = (9)$

Since the sub-tuples are $(2, 2, 2, 3)$, $(2, 2, 2, 5)$, $(2, 2, 2, 7)$, $(2, 2, 2, 11)$, $(2, 2, 3, 5)$, $(2, 2, 3, 7)$, $(2, 2, 3, 11)$, $(2, 2, 5, 5)$, $(2, 2, 5, 7)$, $(2, 2, 5, 11)$, $(2, 2, 7, 11)$.

Again we have given 2 and 3 as first and second digits respectively in the sub-tuples. Thus the question requires the number of combinations taken 4 at a time whose first 2 objects are identified. Then $k_1^* = 1$ and $k_2^* = 4$.

Now in the interval $5 \leq k_3^* \leq$ till 8, we get the assembly of assemblies of selected indexers as

$K_3^* = (K_{31}^*);$ where, $K_{31}^* = (5,8)$

So, $S_3^* = 2$.

Again in the interval $k_3^* + 1 \leq k_4^* \leq$ till 9, we get the assembly of assemblies of selected indexes as

$K_4^* = (K_{41}^*, K_{42}^*)$; where, $K_{41}^* = (6, 8, 9)$

$$K_{42}^* = (9)$$

So, $S_4^* = \sum_{g=1}^{S_3^*} S_{4.g}^*$

$\qquad = S_{41}^* + S_{42}^*$

$\qquad = 3 + 1$

$\qquad = 4.$

Using the summation method we get

$C^* \begin{pmatrix} 9 \\ 4/(2,3) \end{pmatrix} = \sum_{k_3^* \in K_3^*} \sum_{k_4^* \in K_4^*} C$

$\qquad = \sum_{k_4^* \in K_{41}^*} C + \sum_{k_4^* \in K_{42}^*} C$; As $K_3^* = (K_{31}^*)$

$\qquad\qquad\qquad\qquad\qquad\qquad\qquad K_{31}^* = (5, 8)$

$\qquad = 3 + 1$; $\qquad K_{41}^* = (6, 8, 9)$

$\qquad = 4.$ $\qquad\qquad K_{42}^* = (9)$

Since the sub-tuples are (2, 3, 5, 5), (2, 3, 5, 7), (2, 3, 5, 11), and (2, 3, 7, 11).

Corollary 2: The number of combinations of N components that not all different taken V at a time whose first v components are identified, is the same as the number of combinations of $(N - k_v^*)$ components that all different (or not) taken $(V - v)$ at a time i.e.,

$$C^* \begin{pmatrix} N \\ V/{}^v A \end{pmatrix} = C^* \begin{pmatrix} N - k_v^* \\ V - v \end{pmatrix} \qquad\qquad (74)$$

12. Conclusion

From this paper we get idea about combination theorem as well as identified combination theorem. Apart from this, we gain knowledge about the very important general combination theorem.

Acknowledgment

To compose this paper I have taken help from my book 'Bystematics Vol. II My Classic', Scholar's Press, 29 March 2018, ISBN: 978- 620-2-30960-8.

References

1. Bystematics Vol. I My Classic by Deapon Biswas, Scholar's Press 29 March 2018, ISBN: 978- 620-2-30664-5

2. Bystematics Vol. II My Classic by Deapon Biswas, Scholar's Press 29 March 2018, ISBN: 978- 620-2-30960-8.

3. F. Mosteller, R. E. K Rourke & G. B. Thomas Jr. Probability with Statistical Applications.